Secrets Of USB Battery Port Charging

By Dean Billing

I0484166

ISBN 978-1511599436

(Document Version 3.7: 09/08/18)

Table of Contents

Introduction

Back in 2010, I had a conversation with a pilot who needed to charge his iPhone and iPad while flying, from the 12 volt electrical system in his homebuilt aircraft. The iPad would be on during the entire flight providing navigation information. Time in the air could be up to 4 hours in duration and there might be multiple flights during a day on a long cross-country. The electrical system in small airplanes is no different than in a car, so seemed like a simple request.

I knew smart-phones and tablets came with a charger that used the ubiquitous USB cable to connect the device to their wall charger, and the cable was also used to connect said device to a computer for data exchange. Since I'd designed numerous electronic devices, including sophisticated computer peripherals, I thought I could solve his dilemma.

He handed me two Apple chargers, one for his iPhone and one for his iPad-3:

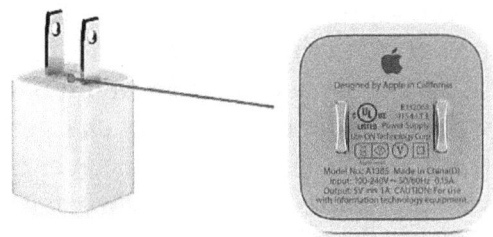

Apple 5 Watt Charger

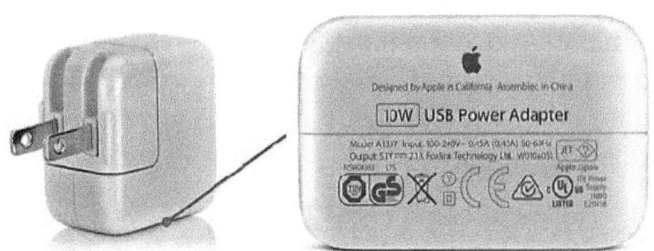

Apple 10 Watt Charger

When I examined his chargers, I thought, "Piece of pie. This will be trivial." After all, the iPad charger was only rated at 5V @ 2.1 A max.

Was I ever wrong!

In my first prototype, I used a simple linear regulator that was able to supply up to 2.5 amps at 5 volts. However, after he installed it in his airplane, he called and said it did not work. Neither his iPad nor his iPhone indicated they were charging.

After I re-tested the device to make sure it worked properly, I began delving into the uncharted world of battery charging using the USB port. The only documentation about this secret realm is supplied by curious minds that have reverse engineered the chargers supplied by manufacturers. References to all of the information I found is in the Acknowledgments And References section of this booklet. None of this information can be found in any manufacturer's documentation that I am aware of. It is not a trade secret, because they use the common USB connector, not some proprietary connector; it is more like some kind of guild secret.

What I learned was that there are two modes for charging the battery in the iPhone, iPad or Android tablet, **Trickle** charging and **Rapid** charging. In order for manufacturers to protect the common USB port in a computer from the power demands of battery charging in newer, more power hungry peripheral devices, they added clever charge management circuitry in the device that limits power draw when it is plugged into a generic computer USB port.

After several design iterations I was able to manufacture chargers suitable for the iPhone, iPad and Android tablets. There are a few companies that make a single charger unit that will *Rapid* charge both an Apple tablet and an Android tablet simultaneously, although I know of one solar charging device and a plethora of car chargers that claim to be able to do it, but don't. All one needs to do is read the negative reviews of the devices to find that out.

Luckily there are new **Smart** integrated circuits that can figure out what type of peripheral device is being plugged into the USB port of a third-party charger and automatically provide the proper signals on the data pins to inform the device it is plugged into a charger that can provide *Rapid* charging capability. There are a few chargers on the market that already incorporate this technology but it is not easy to differentiate them from the myriad of devices that claim to be able to charge any device but cannot.

Warning: All computer USB ports have power limits. Plugging a peripheral device into a port that requires more power than the USB standard specifies may cause damage to the port, possibly resulting in damage to the motherboard. This document contains information about the USB standards for powering peripheral devices.

The documentation for all peripheral devices should clearly state any non-standard power requirements. Please check your documentation carefully before plugging your new toy into a computer USB port, and make sure the USB port on your computer is compatible with your device.

There are now three major USB port specifications and several revisions, and they all have different power and data rate capabilities. Make sure you know which version(s) you have in your computer. The most common one is USB 2, but USB 3 and 3.1 are being included in most new computers. USB 1.x ports are very obsolete but may be in that old desktop or laptop, so verify before you plug something in. USB 1 ports are the most prone to damage from excessive power demand. In fact, the newest USB 3.1 ports can negotiate power availability and if the device being connected doesn't negotiate, the USB 3.1 port limits power to 4.5 watts, or 0.9 amps, which will only *Trickle* charge tablet devices.

Readers would be advised to read the Charging Ports section of the Wiki article and also note that many computers, especially laptops, may shut power off to their USB ports when they are turned off.

1. Charge A Battery On A USB Port

Who's Lame Idea Was That?

There is a misconception that the USB port was originally designed with the capability to charge batteries. Nothing could be further from the truth. Let me state unequivocally: ***The USB Port was NEVER designed to charge batteries.*** Unfortunately, the USB port is a ubiquitous battery charger today because it is a very flexible, inexpensive serial communication specification that just happened to supply some power in its original incarnation. So now, in the latest version of the USB port, there is a specification for battery charging.

Where did the USB port come from? Why was it designed in the first place?

First, let's define what the USB port is. USB is an acronym for the *Universal Serial Bus*. It is: "... an industry standard developed in the mid-1990s that defines the cables, connectors and communications protocols used in a bus for connection, communication, and power supply between computers and electronic devices." You can find this definition and the history of the USB port at the Internet tower of knowledge, Wikipedia. If you peruse the Wiki-babble, you will find just about everything I am going to tell you in this document. However, I am going to distill it, and tailor my remarks to what you need to know to recharge the battery in all of those communication toys you simply can't live without: phones, tablets, game players and navigation devices, etc. Moreover, if you are really a glutton for punishment, you can access the source of the USB specifications at usb.org.

Let's journey back to the 1980s and 1990s and try to discern why the USB port was even invented. The problem solved by the USB specification was the proliferation of low speed PC peripherals that all had different specs and different **connectors**. The mouse had a connector and its own spec. The keyboard had an identical connector, but a different spec. Audio devices, like speakers and microphones, had different connectors and specifications, and then there were printers, some serial, some parallel, and new devices, especially in the media realm like scanners and cameras were proliferating, and people wanted to be able to connect them directly to their computer and download graphic material.

Original IBM PC I/O Connectors

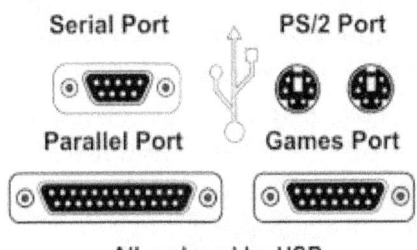

All replaced by USB

The one thing all of these devices had in common was that they could communicate with a computer over a serial interface *if it was fast enough*. And it turns out *fast enough* didn't need to be all that fast back then. So why not define a serial bus specification that was fast enough and flexible enough to satisfy all of these peripheral interface requirements that would use one small physical jack and plug interface. As it often happens, the easy part was agreeing on serial bus technology; the devil was in the details. Standards for the interface were developed in 1994 by seven companies: Compaq, DEC, IBM, Intel, Microsoft, NEC and Nortel.

Note the **absence** of Apple.

What Is A *Serial* Interface?

A serial interface is the epitome of simple electronic connectivity. At a minimum, it only requires three wires to allow transmission and reception. In its simplest incarnation, one wire carries transmit data, one wire carries the receive data, and they share a common ground wire. Everyone could agree on that. In order to make this simple interface attractive to new peripheral designs, which would incur development costs for engineering, they incorporated a fourth wire and put 5 volts on it, like the PS/2 Port, in order to power a low power peripheral device like a mouse or the circuitry in a keyboard. It was rather ingenious, but it also invited abuse of the specification, i.e. to use the port for unintended applications, **like charging a battery**.

Peripheral manufacturers embraced the new specification quickly, and soon keyboards, mice, audio devices, like headsets with microphones, and gaming devices became abundantly available with a USB interface. Interestingly the mouse is essentially a one way device, it only transmits to the computer, but they do require a tiny amount of power for the motion sensors. Keyboards are two way devices: key input goes to the computer, and status lights are driven by the computer, and they do need power for the circuitry that converts key pressings to the codes the computer understands. Obviously the USB specification was a boon, especially as laptop computers emerged where designers did not want to add dedicated connectors for an external mouse, keyboard or game device.

The one additional specification that defines the USB port is the connector. Originally there was only one design, but soon there were a plethora of shapes and sizes, usually smaller as companies created compact peripherals like phones and cameras and wanted miniature connectors. One thing all of the connectors have in common is they have at least four pins for the four wires of the interface, no matter what the size. (Actually, some of the smaller connectors have five pins but they are not germane to this discussion of battery charging.) The USB standards organization decided to designate the connectors with letters. The original connector is known as the USB-A connector and it is the most ubiquitous, but it is also physically the largest:

USB-A Plug

The USB-A plug always has a metal shell, and if you look closely you can see that it has four pins. What you might not notice is that the outside pins are slightly heavier and the two inside pins are closer together. The two inside pins are the transmit and receive data pins, and the outside pins are the ground and power pins. Below is a diagram of the pin position from the socket side:

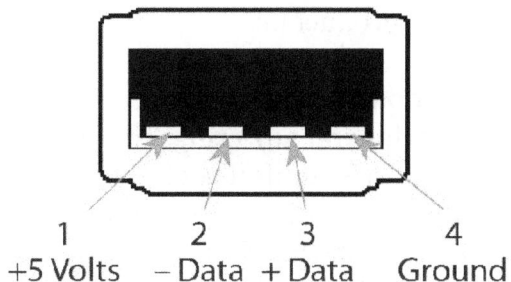

The USB receptacle or *Female* connector

In addition, there are four outer pins around the shell of the receptacle which are ground, thus providing continuity for a cable shield and providing a ground before the voltage and signal pins, which are recessed such that voltage pins engage first. The voltage on pin 1, Vcc, was specified as 5 volts in the USB standard and is the only voltage that should be found on a standard USB 1.x, USB 2, and USB 3 and 3.1, type A socket.

Everything changed with the USB 3.1 standard for non type A sockets, but so far the odd USB 3.1 sockets are only used on an optional 29 watt charger for the iPad Pro 12.9 which is delivered with the Apple 12 watt charger. Therefore, the only socket and voltage we will discuss is the USB-A socket with 5 volts on pin 1.

A note about the USB 1.x, USB 2.0 and USB 3.x standards: information about, and the history of, the standards are discussed in detail on the Wikipedia web page. There were no battery charging standards until 2009 when a *basic charging* protocol was defined: BC1.1 which was updated in 2010 as BC1.2. These battery charging standards were primarily pushed by Chinese and European cell phone manufacturers. Apple had developed their own charger protocol well before 2009 when the iPod 4th generation came out in 2005.

The Basic Charging (BC) standard defines how a *Dedicated Charging Port* (DCP) works by tying the two data lines together with any resistance less than 200 ohms. Usually the charger just shorts the data pins together. A DCP capable USB port can supply up to 1.5 amps, although it doesn't have to; many only supply 1 amp. Few vendors, other than the companies that defined the BC standard follow it, and then not consistently across their product line.

In the USB 3.1 specification there is a detailed, complex specification for battery charging and peripheral powering. The spec was primarily developed for powering peripherals like disk arrays and has not found widespread use for tablet and phone chargers yet. It is being used by Apple to charge batteries in the MacBook series of laptop computers.

2. Voltage / Current / Power

I hate to do this to you, but it is important to discuss a little bit of electronic theory. It is straightforward and the highest math you need is the ability to multiply. Luckily we only have to deal with one voltage: 5 volts. That is the only voltage that a battery charging USB port supplies, so far. However, each new version of the USB standard has delivered slightly more current at the standard 5 volts, thus delivering more power. That was because, as newer peripherals adapted to the USB interface standard, some of them required more power.

In order to deliver more power, the computer had to make more current available at 5 volts. Electrical power is simply voltage times current. Current is expressed in amps and power is expressed in watts. If a peripheral device needs 1 watt at 5 volts, it draws 0.2 amps: 5 volts x 0.2 amps = 1 watt, if my math is correct.

When the original USB 1 specification emerged in January 1996, the committee decided the port need only supply a maximum of 0.15 amps or 0.75 watts. Pretty conservative, but nobody foresaw all of the dandy peripherals we would eventually want. USB 1 and its first improvement USB 1.1 ports were not around very long before peripheral makers were clamoring for more speed and more power. More is always better, right? The USB 2 specification, which came out in April 2000, increased the data rate and increased the max current spec to 0.5 amps, or 2.5 watts. And this is where the technical dam began cracking.

About the time the USB 2 spec was being widely incorporated into computers, cell phone manufacturers were beginning to build *smart-phones*. Smart-phones had a bunch of new features, like camera capability and *Apps,* and had firmware that could be updated. All of these features found their maximum usefulness if the smart-phone could connect to a computer, and the obvious connection medium was a USB interface. An added bonus was the connection could charge the phone while it was plugged into the computer if the charging demand did not exceed 0.5 amps ... by too much. (There is usually a fudge factor built into devices constructed to universal specification.) So the brilliant engineers designed the wall charger for these devices with a USB-A socket for stand alone charging, and the USB cable could be removed from the charger and plugged into a USB-A socket on a computer or laptop to use the serial interface to talk to the computer, while the power interface charged the phone battery. Pretty nifty, huh?

Which brings us to the paradigm shift: if a new peripheral device is so cool that everyone wants it, and it needs to communicate with a computer to fully exploit all of its features to maximum capability, but its battery requires more than 2.5 watts to charge it, how can the USB cable be safely used as the interface?

But before we delve into how certain peripheral manufacturers tackled this conundrum, let me share a couple of observations about battery charging and the power required. Charging a battery is an interesting proposition. In the simplest terms possible let me explain that any 5 volt battery charger for any peripheral device that uses a 5 volt USB charger, cannot be damaged by a charger that is capable of putting out *more* power than is necessary. When the battery in a peripheral device is discharged, it has a slightly lower voltage than when it is fully charged.

In the discharged condition, the greatest voltage differential between the battery and the charger causes maximum current to flow, resulting in the highest power demand. Then, as the battery charges, its voltage rises and its current demand, and thus power demand, decreases. Therefore, the current rating and the power rating on every USB charger you own always specifies the maximum current and maximum power that the charger can supply. So, when charging a battery, power demand is never constant. It is highest at the beginning of the charging cycle, depending on how depleted the battery is, then constantly decreases as the battery charges. And it is the battery that determines the amount of power needed, **not the charger**.

Another thing you should know about battery charging your toys: the battery voltage in the device is **always** less than 5 volts, even fully charged. The common battery voltage for phones and tablets is in the neighborhood of 3.5 volts. Also, there is sophisticated monitoring circuitry to determine when the battery is fully charged, and after it is fully charged, if you leave your device on the charger, it will simply *Trickle* charge to maintain a full charge or disconnect from the charger.

A battery charger must always put out a voltage higher than the battery being charged. Also be advised that even when your peripheral is *OFF*, it is **ON**. When charging, turn your device off so the display is off for quickest charging. The display is the most power hungry component in your device and will prolong the charging cycle if it remains on.

Checking USB Charging Status

There is a way to see how your device is charging. Minimally you need a USB Port volt / ammeter:

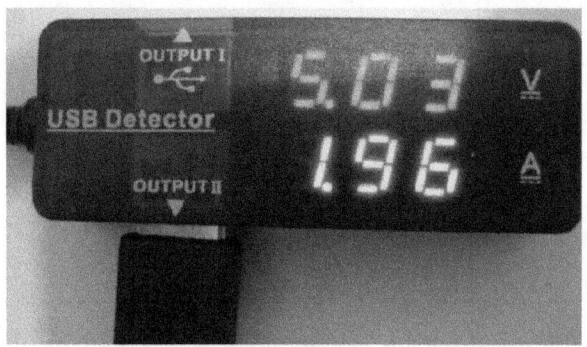

Simple meter displays voltage and current simultaneously

There are also meters that display voltage and current and compute the power:

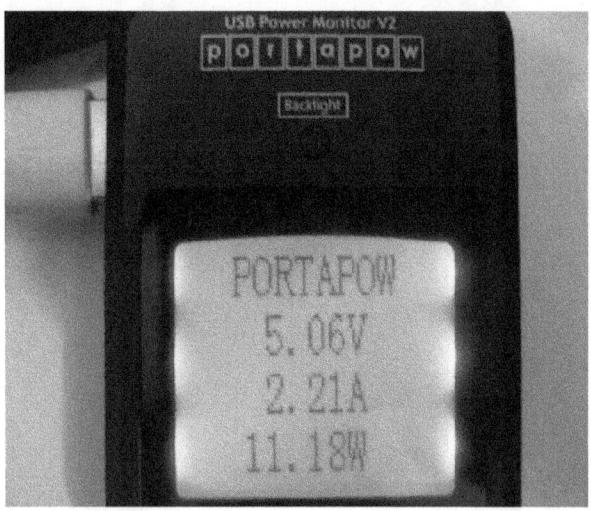

There are a myriad different types of USB meters available at Amazon and on eBay. Most of them are in the $10 range and quality varies widely.

Buyer beware.

There is also a line of sophisticated meters from YZXstudio that will display data pin voltage levels and charging protocols:

YZXstudio USB meter plugged into an Apple 12 watt charger

This intelligent meter examines the voltages on the D+ and D- data pins and informs you which charging protocol it matches on the top two lines of the display.

Warning: Specifications Are Only A Suggestion

It is a good thing that battery charging is a dynamic process. The maximum charging power and highest current demand happens at the beginning of the charge cycle, especially an a completely discharged device. If it did not, there would be a lot of damaged chargers and cables. No USB battery charger is designed for full output, full time. ***It is NOT advisable to use a USB Charger to supply continuous power to a device that needs more than 7.5 watts.*** In other words, it is not a good idea to use your USB charger for any application for which it was not designed.

The USB 2.0 A connector specification has a maximum current rating that the pins can handle of 1.5 amps. One only needs to look at the size of the pins to know that the USB A connector is a low current connector. Therefore, no charger that supplies more than 7.5 watts is within the maximum current specification of any USB 2.0 connector. Yet, the fact is that every 10 to 12+ watt USB charger exceeds the maximum current specification for the USB connector incorporated in the device.

If you read the ratings printed on Apple USB chargers, you will note the 10 watt charger provides current at 5.1 volts, not 5 volts. The USB 1.x and 2.0 maximum voltage spec is 5.25 volts, and luckily, voltage has no affect on the pins in the USB connector unless it is in the hundreds of volts. The Apple 12 watt charger is rated at 2.4 amps maximum, and the output voltage is specified at 5.2 volts; thus, it is actually a 12.48 watt charger, i.e. 5.2 volts x 2.4 amps = 12.48 watts. Obviously, it is hoped the peak charging rate of 2.4 amps will only flow through the connector for a brief period of time since it exceeds the maximum current specification of the connector by 60%.

3. And Then Along Came Apple

The problem started with Apple. Doesn't it always? Everyone who adapted their peripherals, including cell phone makers, to use the USB port to communicate with a computer, also charged the device while connected to the computer. However, power demands increased with each new model, and the Apple iPod 4th generation, in 2005, was the device that finally blew through the USB port power limit of 2.5 watts and required a design gimmick to protect the computer USB port. The iPod 4th generation, and later the iPhone, marketed in 2008, came with a 5 watt charger. The unit was a one inch cube that was clearly labeled as putting out 5 volts at 1 amp max:

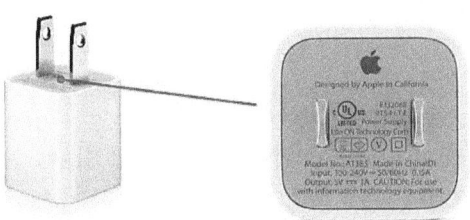

Ubiquitous Apple 5 watt charger

But when the iPad came out, it had a charger that didn't look like any previous charger. It was definitely larger. Up until this time, most cell phone chargers were approximately a one inch cube. The iPad charger looked like a cell phone charger on steroids. When you read the label, it became apparent why.

Every wall wart charger has a label on it that tells you at least two technical parameters that you need to know. It should tell you that it is a 5 volt charger, although some may say 5.1 or even 5.2 volts, and it gives you a current rating, which is the maximum current it will supply. Some chargers may even list the wattage, though that is not necessary since you already know that voltage times current gives you the power available in watts. Below is an example of the verbiage that is on every charger, so look for it:

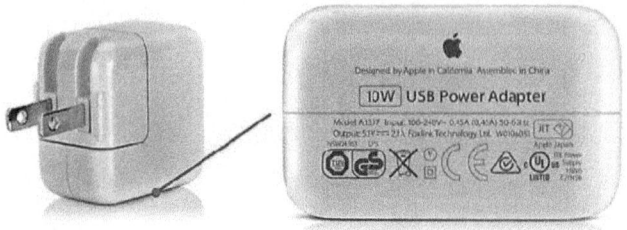

Later iPods, iPhones and iPads needed much more than 0.5 amps to charge a low, or dead, battery, which is the maximum for a USB 2 computer port, and even drew slightly more than the maximum power of the first USB 3 port specification. Therefore, Apple needed a way to limit the charging current when the device was plugged into a computer USB port to use the data interface. They needed a way to protect the computer USB port from damage by drawing too much current when beginning to charge a low battery. Remember, current draw tapers off while recharging the peripheral devices battery, but initially it can approach 1.8 amps for an iPhone and it will require even more for an iPad.

The solution was to engineer two new features in order to enable charging a battery that had a current demand that exceeded the USB 2 and even the maximum USB 3.0 current output specification. They added a provision in the charge monitoring circuitry in the peripheral device to limit current draw for charging when the cable is plugged into a computer USB port and devised a method of sensing whether the cable is plugged into a wall charger or into a computing device. Thus, their peripheral device now has two charging modes, *Rapid* charging when it senses it is plugged into a wall charger and *Trickle* charging when it senses it is plugged into a USB data port on a computing device.

The iPhone and all later Apple devices that use a USB cable for battery charging sense which mode to assume by the signal found on the two data pins of the USB interface. This interface gimmick differentiates those devices that need high power, *Rapid* charging capability, which can only be safely supplied by a wall charger, from those devices that need

0.5 amp or less to recharge the battery, which can safely be supplied by any computer USB 2 interface, or a wall charger. Devices that can recharge their battery within the 0.5 amp maximum current specification of the USB 2 port do not look at the data pins on the interface. Those devices that need more than 0.5 amps to charge a battery perform some gimmick with the data pin signals, but unfortunately, there is **no single standard**. Apple was the first company I am aware of to use the data pins to tell its peripherals what charge rate mode to assume, *Trickle* or *Rapid*.

How An Apple Device Tells If It Is Plugged Into A Charger

A power delivery and battery charging specification is being developed by USB.org. But Apple developed their battery charging solution before any standards were issued. The Apple battery charging solution is described in detail in the Wikipedia article:

"Before the battery charging specification was defined, there was no standardized way for the portable device to inquire how much current was available. For example, Apple's iPod and iPhone chargers indicate the available current by voltages on the D− and D+ lines. When D+ = D− = 2.0 V, the device may pull up to 500 mA. When D+ = 2.0 V and D− = 2.8 V, the device may pull up to 1 A of current. When D+ = 2.8 V and D− = 2.0 V, the device may pull up to 2 A of current."

Apple 5 watt charger protocol

Apple 10 watt charger protocol

What all this means is that any charging device, especially a car charger or a solar charger must have the circuitry necessary, in addition to the power necessary, to supply the voltages as outlined above on the data pins in order to *Rapid* charge a given Apple device. When you plug an Apple device into any USB port that does not have the proper voltages on the data pins, the lightning bolt does not appear, but the device may be *Trickle* charging at 0.5 amps. Unfortunately, Apple devices do not have a separate indicator to inform you they are *Trickle* charging.

Charging indicator, the lightning bolt, on Apple device DOES NOT indicate charge rate

How An Apple Device Charges

There is a new data pin protocol for the Apple 12 Watt charge used with the iPad 4, Air and Pro. A voltage of approximately 2.7 V is presented on both the D+ and D− pins. This charger is backward compatible with earlier iPads and also charges all iPhones and iPods. See the charger compatibility page on the Apple web site.

The battery charging controller inside the iPad 3 and 4, Air, Mini 4, Pro models: 9.7, 10.5 or 12.9 is very intelligent. If it senses that it is plugged into an iPhone 5 watt charger, it will limit charging to 1 amp; If it is plugged into an Apple 10 watt charger, it will charge at a maximum rate of 2.1 amps. When it senses it is plugged into an Apple 12 watt charger, it will *Rapid* charge at a maximum rate up to 2.4 amps. Be advised that the charger lightning bolt icon does not differentiate between the three charging rates. Without a USB volt / ammeter you can't be sure what rate your tablet is charging at.

Apple 12 watt charger protocol

The latest Apple charger is known as the USB-C charger. It first appeared as a 60 watt charger for MacBooks. But now there is a 29 watt version and Apple would have you believe that it is only for the 12" MacBook and the latest iPhones. However, the iPad Pro 10.5 and 12.9 tablets understand the enhanced USB Power Delivery protocol which delivers 14.5V @ 2A, thus 29 watts. The 29 watt USB-C charger has a different type of socket so the old USB-A connector will not plug into it. You will need a special USB-C to Lightning cable for your iPad.

Apple USB-C 29 watt charger on bottom compared to older chargers

The USB-C charger is backward compatible with all iPad tablets in that it can mimick the Apple 12 watt protocol charger. But when you plug an iPad 10.5 or 12.9 into the charger, it negotiates the 29 watt charging protocol. Be advised, if you are using an iPad 12.9 in the cockpit, you will need the USB-C Power Delivery protocol to keep it charged while ON. The old 12 watt protocol cannot keep up with the power demand of this large tablet.

Visual Example Of An Apple Tablet Charging

One rather discharged iPad 4:

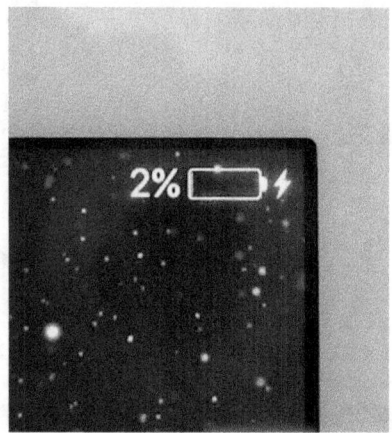

The iPad 4 charging on an Apple 12 watt charger:

12 watt charger before device plugged in

Device charging @ 11.6 watts (2.25 amps)

Notice that the tablet has affected the voltage on the data pins. No matter, it is charging correctly at almost the 12 watt limit of the Apple charger.

What would happen if you plugged your Apple iPad 4, Air or Pro into the older Apple 10 watt charger (2.1 amp max.)?

10 watt charger before device plugged in

Device charging at 10 watt rate (~2 amps)

Finally, how will an iPad 4, Air, or Pro charge on an Apple 5 watt charger (1 amp max.)?

5 watt charger before device plugged in

Device charging at 5 watt rate (<1 amp)

Note: the Apple charge indicator, the lightning bolt, will be on when using any Apple charger, indicating your device is charging, but it does not indicate at what rate your tablet is charging. If an iPad 4, Air or Pro is turned on while plugged into a 10 watt or 5 watt charger, its battery may slowly discharge.

In summary, Apple tablets have smart charging circuitry that is built into the tablet. Also note that late model iPhones are able to sense if they are plugged into an Apple 10 or 12 watt charger and will charge at a higher power, and thus, a faster rate.

There is no charge control circuitry built into a USB-A type wall charger, the Apple 5, 10 or 12 watt models. It will maintain 5 to 5.2 volts and supply as much current as the charging device demands up to its maximum limit: 1 amp for the 5 watt charger, 2.1 amps for the 10 watt charger and 2.4 amps for the 12 watt charger. This does not apply to the

USB-C charger which has two charging modes. One mode emulates the Apple 12 watt protocol which is compatible with all iPads and iPhones, and the new USB Power Delivery protocol which is compatible with the iPad Pro 10.5 and 12.9.

4. Yeah! So What About Everyone Else?

If you read Chapter 3, you know Apple pioneered high power, *Rapid* charging utilizing a USB port in their chargers, even though the generic computer USB port was never designed to supply the amount of power necessary for charging the battery in an iPhone or iPad. Apple had to add circuitry in the device to sense when the USB cable was plugged into a computer, and limit the power draw for battery charging in order to protect the computer electronics.

Other manufacturers, especially those who make smart-phones, have also jumped into the tablet market. So far, I have not run across an Android smart-phone that won't indicate it is charging on a normal USB 2.0 or 3.x computer port. However, check the charger current. The newer smart-phone chargers I have seen have 1+ amp chargers. That power requirement would exceed the limit of the USB 2 and even the USB 3.0 standard, but those phones can plug into computer ports to share data. I have seen no warnings about plugging them into computer USB ports, and my latest Samsung Galaxy S5 indicates it is charging when plugged into my computer. I have verified that it has smart charging capability that limits its load while plugged into a computer.

Android tablets also have 10-12 watt chargers, i.e. they are 5 volt, 2.1+ amp chargers just like the Apple iPad charger. However, there's a key difference. Obviously, they could not use the same scheme that Apple used to signal the need for 10 watts to *Rapid* charge their batteries. I have examined the Samsung Galaxy Tab and the Google Nexus 7 and they have circuitry in the charger tying the two data pins together to a 1.2 volt level to signal the *Rapid* charge capability. I assume this is a standard across Android tablets since I have verified it on two different tablets, but have been unable to locate documentation that there is a standard.

Samsung Android tablet charger protocol

An advantage that some Samsung Android tablets have is that the charging icon has two status indications. It informs the user if it is *Trickle* charging, which happens when it is connected to a computer, or it is *Rapid* charging when connected to a wall charger or a third-party device that understands the signal that must be present on the data pins. When a Samsung Android tablet is *Trickle* charging, a red X appears next to the lightning bolt and there is no animation:

Samsung Android Tablet *Trickle* Charge Icon

When a Samsung Android tablet is *Rapid* charging, the lightning bolt appears and the level in the background is animated and eventually fills the entire battery icon as the battery reaches full charge:

Samsung Android Tablet *Rapid* Charge Icon

5. Types Of USB Battery Chargers

A Frank Discussion About Third-party USB Battery Chargers

You have a device whose battery needs to be charged while you are somewhere other than at home where you can use the wall charger that came with it. You might be in your car, riding your motorcycle, on your yacht, flying your airplane, or in a remote area, off grid. The one thing these non *Original Equipment Manufacturer* (OEM) chargers have in common is that 12 volt, or maybe 24 volt, battery power is available. It is provided by the electrical system of your vehicle or an off grid power supply provided by, say, solar or wind power.

When you purchased the nifty device that you now cannot live without, you may or may not have been offered the option of a *Car Charger* cord, a device that plugs into what used to be called a *Cigarette Lighter.*

Auto Cigarette Lighter Assembly

Most vehicles today have done away with the cigarette lighter element part of the assembly:

but retained the socket, relabeled it a 12 Volt power outlet and generally put a cover on it:

It is one of the few electrical interfaces that is as omnipresent in all mobile vehicles, just as the USB socket is in computers. They are found in cars today as a *Power Outlet* or a *Charging Port*, and they can be found on motorcycles, in most boats as long as they have an electrical system, and in most light aircraft. Just about every device manufacturer offers an optional charging cord that fits in this socket.

However, there is another option, especially if the device manufacturer does not supply a car charger cord. A thousand manufacturers have sprung up to fill the bottomless void of demand for charging devices using the USB port. The problem is, few third-party USB charger vendors understand the *Rapid* vs. *Trickle* charge feature of the Apple and Android products. Thus, their devices only provide power on the USB power pins, 1 and 4, but no signal voltage level on the data pins. Therefore, even though a charger may be rated at 10+ watts, i.e. 5 volts at 2.1 amps or more, they might not *Rapid* charge an iPhone, iPad or Android tablet. Even if the packaging specifically says that the device is made for the iPhone or iPad, that is no guarantee that it will *Rapid* charge said device, and practically no manufacturer knows about *Rapid* charging Android tablets.

Buyer beware.

Very cheap auto store USB charger for your auto "Power Outlet".

Example of misleading packaging. The upper USB socket is not an Apple 10 watt port (2.1 amp). Both ports are Apple 5 watt sockets.

What happens if you buy one of these cute third-party chargers that supposedly puts out 2.1 amps, and you plug it into your mobile *Power Port* but your iPad does not *Rapid* charge. You have two options: return it for a refund and try another brand that has accurate labeling, or buy a *Dongle*.

The latest way to turn any dumb charger into an OEM charger, or to use one OEM charger to charge a competitors device, is to use a *Smart Dongle*. Using the same chip that is found in the new *Smart Chargers*, this gadget will sense what type of device you are plugging into it for charging, and dynamically present the correct voltage signals on the data pins to instruct the device to go ahead and *Rapid* charge, just as if it was plugged into its own wall charger. Here is an example of the Smart Dongles that are coming to market:

Intelligent Smart Charge *Dongle*

Note: Disregard any marketing hype about "Turbo Charging" and **DO NOT** plug this device into a computer or laptop USB port.

Note: It should be obvious that if you must use a Dongle which has been modified to fool your device into thinking it is plugged into its OEM wall charger by putting the proper signal protocol on the data pins, then the gadget cannot transfer data simultaneously to a computing device while charging your toy.

 I am constantly amused by the number of disgruntled customers who bought charging dongles, connected them to a computer USB port and were angry that their computer did not communicate with their toy while *Rapid* charging the device. Of course, the more serious problem they did not understand was, just because they used a *Dongle*, it did not guarantee that the USB port, whether on a third-party charger, laptop or computer, could provide 12+ watts. Certainly, most computer USB ports are not designed to, and the ones that might be able to, which are USB 3.1 ports, must negotiate the power level over the data interface. Without negotiation a USB 3.1 port will not quite supply 5 watts. It would help if manufacturers explained this to their customers. Since they don't, I wrote this book.

Dongles are primarily designed to be used with DUMB third-party chargers made for auto, motorcycle, boat or aircraft use, or off grid charging systems like solar or wind that deliver 12+ volts at 15+ watts. Do NOT use them to charge your toys on your home computer or laptop!

There are smart third party wall chargers. One that works as advertised and *Rapid* charges any device is made by Anker:

Detects your device.
Delivers the fastest charge.

Anker 40W 5-Port USB Desktop Charger
(Knows Apple 12 Watt Protocol)

You can charge up to five different manufacturer's devices and *Rapid* charge them as long as the sum of their current demand doesn't draw more than 40 watts, or about 8 amps.

RFI

There is another, potentially serious, problem you should be aware of with regards to third-party USB battery chargers. This subject is particularly applicable in aircraft use. The acronym RFI stands for Radio Frequency Interference. Every digital device generates RFI, including cell phones, tablets, games, navigation devices, etc. During the design process, steps are taken to minimize RFI so that our toys do not interfere with other devices, especially radio and radio navigation systems, and don't give off physically harmful radiation. The fact is, all digital devices, even non computer devices such as digital cameras, must conform to rigorous RFI limitations defined by the FCC, and you will note that your toys probably have a tag on them somewhere that says they meet FCC part 15 design limitations. Cell phones are regulated under a different set of FCC rules because they contain transmitters which purposefully generate Radio Frequency (RF) energy. The possibility of generating harmful RFI is the principal reason you must turn your toys off when flying commercially below 10,000 ft. to insure that your device does not interfere with the navigation receivers of the aircraft, especially those used in the landing system.

Most third-party charger manufacturers pay little or no attention to RFI suppression since they are not considered a computing device. It is unusual for these devices to interfere with radios in cars and boats, although it is not unheard of according to reviews on Amazon. However, the author knows of numerous instances where these devices interfere with the Communication (COM) receivers in aircraft. There are two major reasons why these devices produce RFI: to make the devices as cheaply as possible, plastic enclosures that provide no shielding are used, and the manufacturer minimizes the parts count by leaving out bypass capacitors which cost money and take up circuit board space, but if present would reduce RFI.

This is a picture of the inside of a common USB charger that plugs into a car charging port:

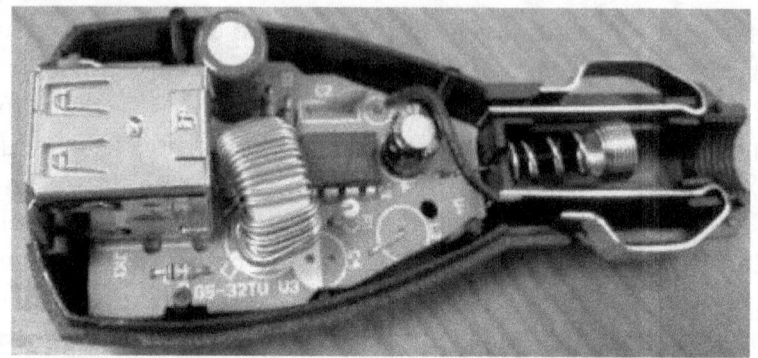

Common Auto USB Battery Charger

Note the unshielded electronics encased in a simple plastic enclosure.

The electronic circuit this device uses to convert 12 volts to 5 volts is called a charge pump. While efficient, it is a terrific RFI generator because of the electrical wave-forms it generates in order to do efficient power conversion. Careful design is a must to reduce the RFI signals.

Heat

All power conversion devices generate heat, especially battery chargers. In the case of the USB port battery chargers, they must convert 120-240 volt AC or 12-24 volt DC to 5 volts DC and supply 2+ amps, or 10+ watts to charge the battery in most modern smart-phone and tablet devices. None of these charging devices is 100% efficient. All modern chargers use a circuit called a *Charge Pump* and most charge pump circuits are 80 to 95% efficient. Ironically the higher efficiency is usually obtained under the heaviest charging load, which is actually a good thing.

Lets look at a worst case analysis of the amount of heat that these small devices must shed during operation. If the device is providing 10 watts to charge the battery in your toy and it is running at 80% efficiency, that means that it is using 12.5 watts from its power source and 2.5 watts is being lost in heat. If that 2.5 watts is being dissipated in a small, plastic cased device plugged into your car charger, which probably does not have much air flowing around it, it will get hot. Considering the number of complaints about these devices in reviews about the device getting hot, you now understand why.

And think about the fact that manufacturers are now offering similar sized products that can provide 4 amps, or 20+ watts. Hopefully, during your battery charging, the peak current demand is short lived and then the power demand tapers off as the battery charges. However, if you hook up a device that demands 10+ watts continuously, the USB port charger will probably get very hot! If you are lucky, it will shut off before melting down. If you are not lucky, it may catch fire. Such failures have been reported in negative reviews on Amazon.

What is ironic is that some OEM wall charging devices get quite hot too, even ones from reputable manufacturers, and some have even caught fire. It is best to monitor all charging devices for excessive heat. Luckily, most of them have circuitry that shuts them down when they overheat, which is one reason that some third-party USB port battery chargers reportedly turn off, then turn on, repeatedly, and drive your charging device crazy. Again, read the reviews of the devices you are contemplating purchasing, especially the negative reviews. There is a treasure trove of information in those reviews.

6. USB Battery Charging Hints

When A Phone Or Tablet Is OFF, It Is Still ON!

For the fastest recharge of a phone or tablet, turn the device off, so the display is off. The display on any device consumes the most power. Remember, parts of the device are still on; especially in a phone, because the connection to the cellular system is always on unless you have put the phone into airplane mode. So, for a faster charge, turn off wi-fi, location (GPS) and Bluetooth, or put it in Airplane mode. Every one of these services consumes power even when the unit is turned off.

For the quickest recharge, power the device down. Be advised it is still consuming a very small amount of power because it has to electrically sense when you press the power button.

I know this is blasphemy, but the charger that came with your device may not be the fastest charger for your toy. For example, all Apple iPhones ship with an Apple 5 watt charger. However, if you charge your late model iPhone on the Apple 10 or 12 watt charger, it may recharge in a much shorter time. The same holds true for your Kindle. The charger that Amazon ships with the Kindle is a cheap 5 watt charger. However, if you recharge your Kindle on an Apple or Samsung 10 or 12 watt charger, it will recharge quicker because the charging circuitry in the device determines how much power it needs to recharge the batteries. And no, you cannot damage the Kindle by charging it on a more powerful charger.

If you have an Apple iPad that came with an Apple 12 watt charger, use the charger to charge all of your Apple gadgets. All of them can sense which Apple charger they are plugged into and charge accordingly. Actually, the Apple 10 watt charger that came with the early iPads through the iPad 3, the iPad Mini 2, 3, 4, the iPad Air 2 and the iPad Pro 9.7, is no longer available for replacement at the Apple store, but your early iPads can be charged on the Apple 12 watt charger or any *intelligent* 12 watt third party charger. Ironically, the iPad 3, Mini 4, Air 2 and Pro 9.7 will charge faster on the 12 watt charger.

Want to minimize the number of chargers when you travel with a myriad of devices by different manufacturers? Take any 12 watt charger and an intelligent dongle:

A 12 watt charger and a *Smart Dongle*

USB Chargers At Airports, In Airplanes And In Hotel Rooms

Most are not intelligent, although I found one intelligent charger port in a brand new hotel on a recent trip. You will be lucky if they present the Apple 5 watt protocol, but will probably present the computer USB 2.0 protocol of 5 volt / 0.5 amp = 2.5 watts. Thus, using an intelligent dongle won't make any difference with the airport and hotel room chargers, although, it doesn't hurt to give it a try. Worst case the voltage will drop, and your device won't charge at all.

Be aware that there are many new, oddball, proprietary charging protocols, but they are backward compatible with all 5 volt chargers. Examples are the Qualcomm Quickcharge 2.0 and the Samsung Adaptive Fast Charging: 9V @ 1.6 amp = ~14 watts. When you see one of these charging protocols on your OEM charger, you will note they all have current limitations under 2 amps. The obvious reason for that is the current specifications for contacts in most USB connectors is less than 2 amps.

Tablets and phones that communicate over USB ports to computers will usually trickle charge when connected to any computer USB 2.0 port even though they may display a message that they are not charging, or don't display the battery charging icon. Some devices will charge at about a 1.4 amp rate on a USB 3 port and may even indicate they are charging. Charging a device on a computer USB port is an emergency method, but you will need a lot of time to go from a discharged battery to full charge.

Hints For Flying With Your iPad

Know your charging system before you have to depend on it

Since this booklet came about because of a requirement for charging an iPad in an airplane, here are some hints for flying with your iPad and iPhone.

Normally, you probably turn your tablet off to charge it, but you might not be able to do that while flying an airplane. You definitely don't want the device to completely discharge during flight. Not only do you want it to at least maintain its charge level, more than likely you want it to charge up to full charge while on. To do this with the iPad 4, Air or Pro 9.7, you have to have an on board charger that knows the Apple 12 watt charging protocol. It also makes sense to fully charge your device before you fly, if possible. Without a 12 watt charging system in the aircraft, a late model iPad may discharge while on and plugged into a charger.

iPad 4 charging while at 50% depletion: 11.4 watts needed

If your iPad is discharging while turned on and plugged into a charger, the charger probably doesn't understand the Apple 12 watt protocol. More than likely it only knows the 10 watt protocol (2.1 amp) at best.

Flying with a discharging iPad plugged into a cheap auto converter will probably cause the converter to run hot and it may intermittently cut out, and the iPad will slowly discharge.

Have A Charging Problem?

Suspect The Cable

Something has changed; your iPad is discharging while you are flying. It is plugged into the appropriate Apple 12 watt or 10 watt capable charger. What do you suspect, the charger, the iPad, or the cable between them?

The iPad Lightning cable is an active device, that is, it has embedded electronics. The electronic parts are in the connector that plugs into the tablet:

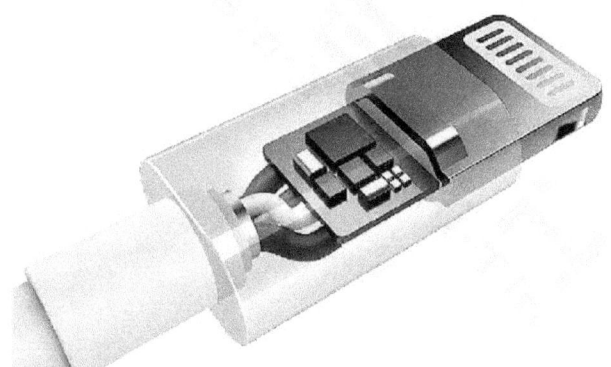

First, you should disconnect your charger cable **at your iPad**. Wait a few seconds and plug it back in. Every so often, I have seen iPads fail to sense the charger protocol and fall back to the 5 watt charging rate which is only 1 amp, or worse, the computer USB port rate of 500 ma.

Next, you need to verify the charger is presenting the Apple 12 watt protocol, or the 10 watt protocol if that is all your older iPad recognizes. This is where you need a USB volt / ammeter that can show you the USB data pin voltages as well as the power supplied by the charger when charging your iPad.

Once you have verified the charger is indeed acting as the appropriate Apple charger, you should suspect the cable, especially if it is a 3rd party cable. There are a few 3rd party cables that are better than Apple cables, but the majority of them aren't. You should always carry a spare Apple Lightning cable with you on flights and swap out a suspect cable.

Unfortunately, cables have two failure modes. The first is easy to diagnose: a total failure because of broken wires in the cable, usually at the connectors. The symptom is the lightning bolt charge indicator doesn't come on. The other failure mode is more insidious. The lightning bolt charge indicator is on, but your tablet isn't charging properly. This is another reason why you need a USB volt / ammeter. I have seen both an earlier Apple 30 pin charger cable and an Apple Lightning cable that reduce the current supplied to an iPad significantly. Replacing the cable resolved the problem in both cases; the suspect cables were both Apple cables.

A final note about the Apple Lightning cable, let me reiterate: The Lightning cable is not a passive device only made up of wires and connectors. It is an active electronic bus as described in Wikipedia. There is an I.C. chip in the Lightning connector, thus presenting another failure mode in the cable assembly.

Beware Counterfeit Chargers

Don't Believe The Packaging Or The Label

While Apple no longer sells the USB 10 Watt Power Adapter in the Apple store, that doesn't mean there aren't millions of them around. Turns out Apple still ships the 10 watt charger with the iPad Mini 4, and shipped it with the Air 2 and early Pro 9.7, even though each of those iPads will charge faster on the 12 watt charger.

Turns out though, there may be thousands of 10 watt chargers, genuine and knock-offs, that are being sold as 12 watt chargers.

Example of fake Apple 12 Watt Power Adapter

I purchased the charger pictured above on eBay. The packaging was genuine, at least it looked genuine, and the unit was stamped with the Apple 12 watt charger verbiage. It is not the only 12 watt charger I purchased on eBay that turned out to be a 10 watt charger of unknown origin, but it is the best example of what is being passed off as genuine, new, OEM Apple 12 watt chargers.

Plugging a YZXstudio volt / ammeter into the charger shows that it is actually a 10 watt charger, even though the stamp on the charger case says it is an Apple 12 watt charger:

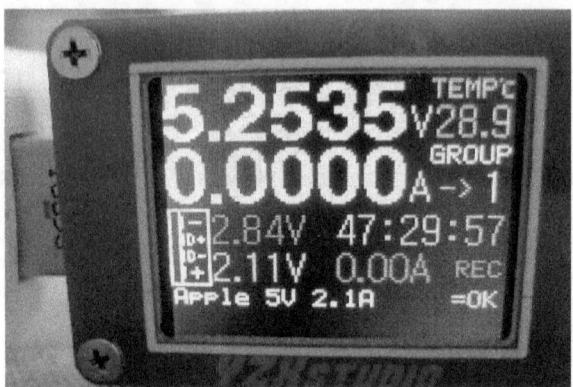

Definitely a 10 Watt Power Adapter

When I used the power adapter to charge an iPad Air 2 that was at 15% charge, it supplied just over 2 amps as you would expect a 10 watt charger to supply. It did not supply 2.3 - 2.4 amps that a 12 watt charger would put out, and it got quite warm.

Charging iPad Air 2 at 10 watt rate

Always check every Apple Power Adapter you purchase, even from Apple, to verify that it is marked correctly and has the proper protocol voltages on the data pins. Had the above adapter been a 12 watt charger it would have had approximately 2.7 volts on both data pins and supplied 2.4 amps to my discharged iPad Air 2.

And don't expect the sellers to know what they are selling. I asked another eBay vendor if they verified that the OEM Apple 12 watt Power Adapters were really 12 watt chargers and not 10 watt chargers. Here is their response:

"Yes. We've been selling these chargers for years and can with confidence [sic] they are 12W (2.1A) chargers and not 10W (2A)."

I pointed out to the vendor that there was no Apple 2A charger and the 2.1A charger was the 10 watt charger, not the 12 watt charger, which would supply 2.4A and is labeled as such. Even the fake ones.

Buyer beware.

Appendix

Apple iPad Charging Matrix

Most of us use our iPad tablet to surf the web, check our eMail, text and Tweet, do FaceTime and play games. When the battery gets low, we usually turn our device off and plug it into a wall charger. However, when tablets are used by pilots, the tablet is on continuously during flight and the user would prefer that the batteries remain charged. The pilot may also rely on the tablet to perform flight planning and may arrive at the aircraft with a discharged tablet. It is important that the tablet can charge during flight.

The following Apple iPads were tested for their recharging characteristics while ON, as they would be in the cockpit.

Notes: These measurements were taken on iPads with a variety of hardware configurations and the results may not match your particular iPad that has a different hardware configuration while running your favorite software. All my tests were run with Bluetooth on, Wi-Fi on, GPS on and wireless on if the iPad had it. The one software App always running was Flight Aware since it uses the GPS and Wi-Fi or Wireless to download the flight information and terrain maps.

The iPad Mini comes with a 5 watt charger but will charge faster on the Apple 10 watt or 12 watt charger. Four iPad models: the Mini 4, iPad 3, Air 2 and the Pro 9.7, come with 10 watt chargers, but recognize the Apple 12 watt charger protocol and will charge faster if they are plugged into the Apple 12 watt charger. This is important when using these iPads in the cockpit where they are on for long periods of time.

iPad Mini

CPU: A5
CPU Speed: 1 GHz
CPU Architecture: 32 Bit
CPU Cores: 2
RAM: 512 MB
Storage: 16 GB
Wireless Option: No
Charger Supplied: 5 watt
This iPad recognizes 10 and 12 watt chargers, and will charge faster on them.

Charging Characteristics While ON:

5 Watt Charger:
Battery dead: 1A / 5W, Turns on @2%, charging
Battery 50%: 1A / 5W, charging
Battery 95%: 1A / 5W, charging
Battery 100%: .6A / 3W, maintaining full charge

10 Watt Charger:
Battery dead: 1.8A / 9W, Turns on @2%, charging
Battery 50%: 1.8A / 9W, charging
Battery 95%: 1.3A / 6.5W, charging
Battery 100%: .6A / 3W, maintaining full charge

12 Watt Charger:
Battery dead: 1.8A / 9W, Turns on @2%, charging
Battery 50%: 1.8A / 9W, charging
Battery 95%: 1.3A / 6.5W, charging
Battery 100%: .6A / 3W, maintaining full charge

iPad Mini 2

CPU: A7
CPU Speed: 1.3 GHz
CPU Architecture: 64 Bit
CPU Cores: 2
RAM: 1 GB
Storage: 64 GB
Wireless Option: Yes
Charger Supplied: 10 watt
This iPad recognizes 12 watt charger, but charges at 10 watt rate.

Charging Characteristics While ON:

5 Watt Charger:
Battery dead: 1A / 5W, Turns on @2%, starts discharging
Battery 50%: 1A / 5W, discharging
Battery 95%: 1A / 5W, discharging
Battery 100%: 1A / 5W, discharging

10 Watt Charger:
Battery dead: 2A / 10W, Turns on @2%, charging
Battery 50%: 2A / 10W, charging
Battery 95%: 1.6A / 8W, charging
Battery 100%: .9A / 4.5W, maintaining full charge

12 Watt Charger:
Battery dead: 2A / 10W, Turns on @2%, charging
Battery 50%: 2A / 10W, charging
Battery 95%: 1.6A / 8W, charging
Battery 100%: .9A / 4.5W, maintaining full charge

iPad Mini 3

CPU: A7
CPU Speed: 1.3 GHz
CPU Architecture: 64 Bit
CPU Cores: 2
RAM: 1 GB
Storage: 16 GB
Wireless Option: No
Charger Supplied: 10 watt
This iPad recognizes 12 watt charger, but charges at 10 watt rate.

Charging Characteristics While ON:

5 Watt Charger:
 Battery dead: 1A / 5W, Turns on @2%, charging slowly
 Battery 50%: 1A / 5W, charging slowly
 Battery 95%: 1A / 5W, charging slowly
 Battery 100%: .6A / 3W, maintaining full charge

10 Watt Charger:
 Battery dead: 2A / 10W, Turns on @2%, charging
 Battery 50%: 2A / 10W, charging
 Battery 95%: 1.4A / 6W, charging
 Battery 100%: .6A / 3W, maintaining full charge

12 Watt Charger:
 Battery dead: 2A / 10W, Turns on @2%, charging
 Battery 50%: 2A / 10W, charging
 Battery 95%: 1.6A / 8W, charging
 Battery 100%: .6A / 3W, maintaining full charge

iPad Mini 4

CPU: A8
CPU Speed: 1.5 GHz
CPU Architecture: 64 Bit
CPU Cores: 2
RAM: 2 GB
Storage: 16 GB
Wireless Option: Yes
Charger Supplied: 10 watt
This iPad recognizes 12 watt charger, and will charge faster on it.

Charging Characteristics While ON:

5 Watt Charger:
Battery dead: 1A / 5W, Turns on @2%, charging
Battery 50%: 1A / 5W, charging
Battery 95%: 1A / 5W, charging
Battery 100%: 1A / 5W, maintaining full charge

10 Watt Charger:
Battery dead: 2.1A / 10.5W, Turns on @2%, charging
Battery 50%: 2A / 10W, charging
Battery 95%: 1.8A / 9W, charging
Battery 100%: .8A / 4W, maintaining full charge

12 Watt Charger:
Battery dead: 2.3A / 11.5W, Turns on @2%, charging
Battery 50%: 2.2A / 11W, charging
Battery 95%: 1.5A / 7.5W, charging
Battery 100%: .7A / 3.5W, maintaining full charge

iPad

CPU: A4
CPU Speed: 1 GHz
CPU Architecture: 32 Bit
CPU Cores: 1
RAM: 256 MB
Storage: 64 GB
Wireless Option: Yes
Charger Supplied: 10 watt
This iPad recognizes 12 watt charger, but charges at the 10 watt rate.

Charging Characteristics While ON:

5 Watt Charger:
Battery dead: 1A / 5W, Turns on @2%, charging slowly
Battery 50%: 1A / 5W, charging slowly
Battery 95%: 1A / 5W, charging slowly
Battery 100%: .8A / 4W, maintaining full charge

10 Watt Charger:
Battery dead: 2A / 10W, Turns on @2%, charging
Battery 50%: 1.8A / 9W, charging
Battery 95%: 1.5A / 7.5W, charging
Battery 100%: .8A / 4W, maintaining full charge

12 Watt Charger:
Battery dead: 2A / 10W, Turns on @2%, charging
Battery 50%: 1.8A / 9W, charging
Battery 95%: 1.6A / 8W, charging
Battery 100%: .8A / 4W, maintaining full charge

iPad 2

CPU: A5
CPU Speed: 1 GHz
CPU Architecture: 32 Bit
CPU Cores: 2
RAM: 512 MB
Storage: 64 GB
Wireless Option: Yes
Charger Supplied: 10 watt
This iPad recognizes 12 watt charger, but charges at the 10 watt rate.

Charging Characteristics While ON:

5 Watt Charger:
Battery dead: 1A / 5W, Turns on @2%, barely maintains charge
Battery 50%: 1A / 5W, charging slowly
Battery 95%: 1A / 5W, charging slowly
Battery 100%: .9A / 4.5W, maintaining full charge

10 Watt Charger:
Battery dead: 2A / 10W, Turns on @2%, charging
Battery 50%: 2A / 10W, charging
Battery 95%: 1.6A / 8W, charging
Battery 100%: .9A / 4.5W, maintaining full charge

12 Watt Charger:
Battery dead: 2A / 10W, Turns on @2%, charging
Battery 50%: 2A / 10W, charging
Battery 95%: 1.7A / 8.5W, charging
Battery 100%: .9A / 4.5W, maintaining full charge

iPad 3

CPU: A5X
CPU Speed: 1 GHz
CPU Architecture: 32 Bit
CPU Cores: 2
RAM: 1 GB
Storage: 16 GB
Wireless Option: No
Charger Supplied: 10 watt
This iPad recognizes 12 watt charger, and will charge faster on it.

Charging Characteristics While ON:

5 Watt Charger:
Battery dead: 1A / 5W, Turns on @2%, starts discharging
Battery 50%: 1A / 5W, maintaining
Battery 95%: 1A / 5W, charging slowly
Battery 100%: .7A / 3.5W, maintaining full charge

10 Watt Charger:
Battery dead: 2.1A / 10.5W, Turns on @2%, charging
Battery 50%: 2.1A / 10.5W, charging
Battery 95%: 1.7A / 8.5W, charging
Battery 100%: .8A / 4W, maintaining full charge

12 Watt Charger:
Battery dead: 2.4A / 12W, Turns on @2%, charging
Battery 50%: 2.3A / 11.5W, charging
Battery 95%: 1.8A / 9W, charging
Battery 100%: .8A / 4W, maintaining full charge

iPad 4

CPU: A6
CPU Speed: 1.4 GHz
CPU Architecture: 32 Bit
CPU Cores: 2
RAM: 1 GB
Storage: 16 GB
Wireless Option: No
Charger Supplied: 12 watt

Charging Characteristics While ON:

5 Watt Charger:
Battery dead: 1A / 5W, Turns on @2%, starts discharging
Battery 50%: 1A / 5W, discharging
Battery 95%: 1A / 5W, discharging
Battery 100%: .6A / 3W, discharging

10 Watt Charger:
Battery dead: 2A / 10W, Turns on @2%, slowly charging
Battery 50%: 2A / 10W, slowly charging
Battery 95%: 1.9A / 9.5W, charging
Battery 100%: .9A / 4.5W, maintaining full charge

12 Watt Charger:
Battery dead: 2.2A / 11W, Turns on @2%, charging
Battery 50%: 2.2A / 11W, charging
Battery 95%: 2A / 10W, charging
Battery 100%: .8A / 4W, maintaining full charge

iPad Air

CPU: A7
CPU Speed: 1.4 GHz
CPU Architecture: 64 Bit
CPU Cores: 2
RAM: 1 GB
Storage: 32 GB
Wireless Option: Yes
Charger Supplied: 12 watt

Charging Characteristics While ON:

5 Watt Charger:
Battery dead: 1A / 5W, Turns on @2%, starts discharging
Battery 50%: 1A / 5W, discharging
Battery 95%: 1A / 5W, discharging
Battery 100%: 1A / 5W, discharging

10 Watt Charger:
Battery dead: 2A / 10W, Turns on @2%, slowly charging
Battery 50%: 2A / 10W, slowly charging
Battery 95%: 1.9A / 9.5W, slowly charging
Battery 100%: 1.7A / 8.5W, maintaining full charge

12 Watt Charger:
Battery dead: 2.3A / 11.5W, Turns on @2%, charging
Battery 50%: 2.3A / 11.5W, charging
Battery 95%: 1.7A / 8.5W, charging
Battery 100%: 1.7A / 8.5W, maintaining full charge

iPad Air 2

CPU: A8X
CPU Speed: 1.5 GHz
CPU Architecture: 64 Bit
CPU Cores: 3
RAM: 2 GB
Storage: 64 GB
Wireless Option: Yes
Charger Supplied: 10 watt
This iPad recognizes 12 watt charger, and will charge faster on it.

Charging Characteristics While ON:

5 Watt Charger:
Battery dead: 1A / 5W, Turns on @2%, start very slow charging
Battery 50%: 1A / 5W, barely charging
Battery 95%: 1A / 5W, barely charging
Battery 100%: 1A / 5W, Maintain full charge barely, intermittent discharge

10 Watt Charger:
Battery dead: 2.1A / 10.5W, Turns on @2%, charging
Battery 50%: 2A / 10W, charging
Battery 95%: 1.9A / 9.5W, charging
Battery 100%: ~1A / 5W, maintaining full charge

12 Watt Charger:
Battery dead: 2.4A / 12W, Turns on @2%, charging
Battery 50%: 2.4A / 12W, charging
Battery 95%: 2A / 10W, charging
Battery 100%: ~1A / 5W, maintaining full charge

iPad Pro 9.7

CPU: A9X
CPU Speed: 2.16 GHz
CPU Architecture: 64 Bit
CPU Cores: 2
RAM: 2 GB
Storage: 32 GB
Wireless Option: Yes
Charger Supplied: 10 watt
This iPad recognizes 12 watt charger, and will charge faster on it.

Charging Characteristics While ON:

5 Watt Charger:
Battery dead: 1A / 5W, Turns on @2%, starts discharging
Battery 50%: 1A / 5W, slowly discharging
Battery 95%: 1A / 5W, slowly discharging
Battery 100%: 1A / W, slowly discharging

10 Watt Charger:
Battery dead: 2A / 10W, Turns on @2%, charging
Battery 50%: 2A / 10W, charging
Battery 95%: 1.6A / 8W, charging
Battery 100%: ~1.5A / 7.5W, maintaining full charge

12 Watt Charger:
Battery dead: 2.3A / 11.5W, Turns on @2%, charging
Battery 50%: 2.3A / 11.5W, charging
Battery 95%: 1.7A / 8.5W, charging
Battery 100%: ~1.5A / 7.5W, maintaining full charge

Newer iPad Models Not Tested

New iPad Pro 9.7 versions are iPad "Year" models. iPad Pro 9.7 discontinued in 2017.

iPad 2017 (Slower than iPad Pro 9.7, but cheaper):

CPU: A9
CPU Speed: 1.85 GHz
CPU Architecture: 64 Bit
CPU Cores: 2
RAM: 2 GB
Storage: 32 GB or 128 GB
Charger Supplied: 12 watt

iPad 2018:

CPU: A10
CPU Speed: 2.3 GHz
CPU Architecture: 64 Bit
CPU Cores: 4, only used 2
RAM: 2 GB
Storage: 32 GB or 128 GB
Charger Supplied: 12 watt

iPad Pro 10.5:

CPU: A10X
CPU Speed: 2.39 GHz
CPU Architecture: 64 Bit
CPU Cores: 6
RAM: 4 GB
Storage: 64 GB, 256 GB or 512 GB
Charger Supplied: 12 watt (Recognizes 29 watt USB-C charger: USB Power Delivery - 14.5V @2A)

iPad Pro 12.9 (Two generations):

CPU: Gen 1: A9X / Gen 2: A10X
CPU Speed: Gen1: 2.26 GHz / Gen 2: 2.39 GHz
CPU Architecture: 64 Bit
CPU Cores: Gen 1: 2 / Gen 2: 6
RAM: 4 GB
Storage: 64 GB, 128 GB, 256 GB or 512 GB
Charger Supplied: 12 watt (Recognizes 29 watt USB-C charger: USB Power Delivery - 14.5V @2A)
Note: iPad Pro 12 needs 29 watt charger to remain charged while ON.

Acknowledgments

Wikimedia Commons for permission to use short quotes and graphics.

References With Links

Choosing USB Pin Voltages for iPhones and iPads – Update for iPhone 5 by Phillip Stearns is an introductory discussion of the data pin voltages that signal USB charger power availability.
[https://www.voltaicsystems.com/blog/choosing-usb-pin-voltages-for-iphones-and-ipads/]

This article discusses the circuitry to implement the iPhone and iPad 5 watt and 10 watt charger protocols showing the value of the resistor ladders that present the proper voltages on the data pins to signal USB charger power.
[https://forums.adafruit.com/viewtopic.php?f=15&t=16729]

This forum topic discusses circuitry to implement an Android charger.
[https://forum.xda-developers.com/showthread.php?t=975558]

This article discusses circuitry to implement several types of USB chargers including Motorola Droid, Samsung devices and an iPhone and iPad *Trickle* charger which would also work on most smart-phones and *Trickle* charge most tablets.
[http://www.obddiag.net/usb-power.html]

Suntactics, is an example of a company that makes third-party USB charging devices for off premise recharging of batteries for your Apple toys.
[https://www.suntactics.com/]

If you are interested in a dissection and analysis of an Apple iPhone charger, this article by Ken Shirriff is an excellent example of how Apple chargers were taken apart and analyzed to provide the documentation outlined in some previous links. It includes copious notes and links to other articles including an eye opening article dissecting an Apple knock-

off charger and some of the dangers inherent is using them. Ironically even Apple has had problems with their USB charger designs. The article describes the embarrassing recall in 2008 of iPhone chargers "... due to a defect that the AC prongs could fall off the charger and get stuck in an outlet."
[http://www.righto.com/2012/05/apple-iphone-charger-teardown-quality.html]
[http://www.righto.com/2012/03/inside-cheap-phone-charger-and-why-you.html]

Mr. Shirriff also did a dissection of the iPad charger and a dangerous knock off. If you take the time to read this article, please pay particular attention to the explanation about *noise* and the pretty oscilloscope pictures. This is an excellent illustration of why many third-party USB chargers emit RFI. The output of the charger should be DC with no AC riding on it, especially high frequency AC. An interesting result of all that noise is that it not only may interfere with radios in the vicinity but it apparently can drive your touchscreen crazy.
[http://www.righto.com/2014/05/a-look-inside-ipad-chargers-pricey.html]